Serpents of Knowledge

Written by Nikola Tucakov

Second Edition
Copyright © 2025 Nikola Tucakov

ISBN 978-1-7382000-0-9 Paperback
ISBN 978-1-7382000-1-6 Electronic Book

www.serpentsofknowledge.com

BECOMING US

Life always adapts to changes in its environment.

A little black frog of Maui, by means of natural selection, managed to survive on a specific black-patch of land created by lava flow. A drastic and almost instant change in this frog's evolution.

Sometimes the environment changes slowly, like the slow-sinking islands of Galapagos. Over millions of years, the same species of animals and plants develop slight changes across different islands, branching them all towards their new evolutionary paths.

In the same fashion, there is a string of events in our past that made us humans that we are today.

About forty-million years ago, India collided with Asia creating the rapid uplift of the Himalayas.

This newly-formed barrier initiated the monsoons and inadvertently dried-out all of North Africa that was at the time covered in thick rain-forests. The inhabiting primates that depended on the trees had to quickly adapt to this new dry-land environment, and thus they started to walk.

The land didn't completely dry however. Roughly every thirty-thousand years, North Africa experiences periods of high-flooding, turning Sahara lush and green.

Our common ancestor, Lucy the ape, who lived in Africa just over three-million years ago, was able to walk as well as climb trees. Exactly what this kind of environment would produce.

During the dry periods, the primates would migrate out of North Africa, settling in various other parts of the world.

Following this colossal continental collision in South Asia, Earth's climate started gradually cooling and around 34 million years ago, for the first time in more than 200 million years of warm-global-weather, the Southern Antarctic ice-cap was formed.

Then, only about 2.6 million years ago, when the American continents joined at Panama, separating the Pacific and the Atlantic Ocean, the Gulf stream was created, freezing the North, and so creating the second, Arctic ice-cap.

Large glacial-masses expanded into the vast northern-lands as sea-level dropped by hundreds of feet, constantly shifting the coastlines. Bound by two glacial masses, Earth entered a state of extreme environmental change.

The glacial cycles.

Earth's climate was now changing much more dramatically than what life on this planet had been used to for a very long period of time. This particularly affected these walking primates.

To survive, the hominids had to constantly migrate, as their land would eventually be rendered unliveable.

It is during this time that the ape's brain grew twice in size. Their bodies got larger, guts, jaws and teeth grew smaller, and bodily-hair got thinner. All consistent with them harnessing fire.

Eventually, they figured-out how to build shelters, tools and clothing, essentially adapting themselves to the environment, and thus so migrating less. They carried this newly-acquired knowledge in their newly-evolved brains.

This very challenging and fast-changing new environment that started just 2.6 million years ago, by means of natural selection, gave way to the emergence of not only the Homo Sapiens, but many other smart bipedal hominid species.

Our brain is the product of the environmental change linked to the start of the glacial cycles.

The concept of a punctuated equilibrium in evolution says that fossils seem to stay in similar evolutionary states for long periods of time. Then suddenly, often after cataclysms, they change into a different evolutionary equilibrium. This new stable design then remains bound by the new stable environmental pattern until another change comes along.

The "monkey to human" transition seems to have gone through three principal equilibrium changes.

When India hit Asia some 40 to 30 million years ago, the monkeys in North Africa evolved into a new "walking equilibrium". These primates were the Australopithecus (Lucy), Homo Habilis and Rudolfensis among many, and even though they developed the ability to walk, they still depended on the trees.

When the glacial cycles arrived 2.6 million years ago, forcing some of these primates to constantly migrate, the second "bipedal equilibrium" was reached. This is the actual point when we left the trees and we started walking. The prime example is the Homo Erectus, with many others like the Homo Ergaster, Antecessor and Heidelbergensis. Even though these primates used tools and harnessed fire, to us, they would still have come across as animals.

The third, and the last equilibrium in our evolution came with the building of a residence, a home. The primates that decided to resist the constant migration started building permanent shelters, that were socially very contained environments. This instigated the development of higher social and cognitive skills, creating the new "smart equilibrium" with primates like the Homo Sapiens as well as the Denisovans and Neanderthals, both of whom had larger brains than us. The oldest discovered fossil is Homo Longi, identified in 2025 as Denisovan, dated back to 1.1 million years old.

This last cognitive equilibrium, that is the growth of our neocortex, seems to correspond to the second-order theory of the mind. A recursive way of thinking that evokes sarcasm, imagination and most importantly understanding that someone's beliefs can be influenced by a third person. It creates a world with large and complex social structures, and is essentially what sets us apart from animals. Animals possess the first-order theory of the mind, understanding what others believe, and can experience things like empathy and deception. This evolutionary change also matches a genetic alteration in our chromosomes. The Great Apes had 24 pairs of chromosomes until the Homo Erectus equilibrium, after which the DNA fused into 23 pairs that we the humans, as well as all the hominids in the last equilibrium have (Neanderthals and Denisovans).

The difference between a human and an animal home is that animals can change the environment of their habitat, but a human can actually control it.

This is most easily done with fire, and the biggest challenge in handling fire is smoke. That's why the caves were ancient people were discovered were of particular shapes that could handle smoke.
Though caves seem more of a place of refuge than habitat, if people were smart enough to choose the shape of the cave, they would also understand the concept of a chimney, which would technically be the first human invention. The first human home, that created the human neocortex, the fusion of the chromosomes in 23 pairs and the development of the second order theory of mind. All of this would have happened at least 1.1 million years ago.

When time is magnified like this, it makes our current civilization look very short. Seems like our transformation into these "smart beings" happened in just mere few-thousand years of our history.

But in reality, this has been happening for hundreds of thousands of years. Our ancient relatives were nothing short of how smart we are today, with some quite possibly even smarter.

We have inherited a very sophisticated brain; it is questionable whether we actually know how to use it properly.

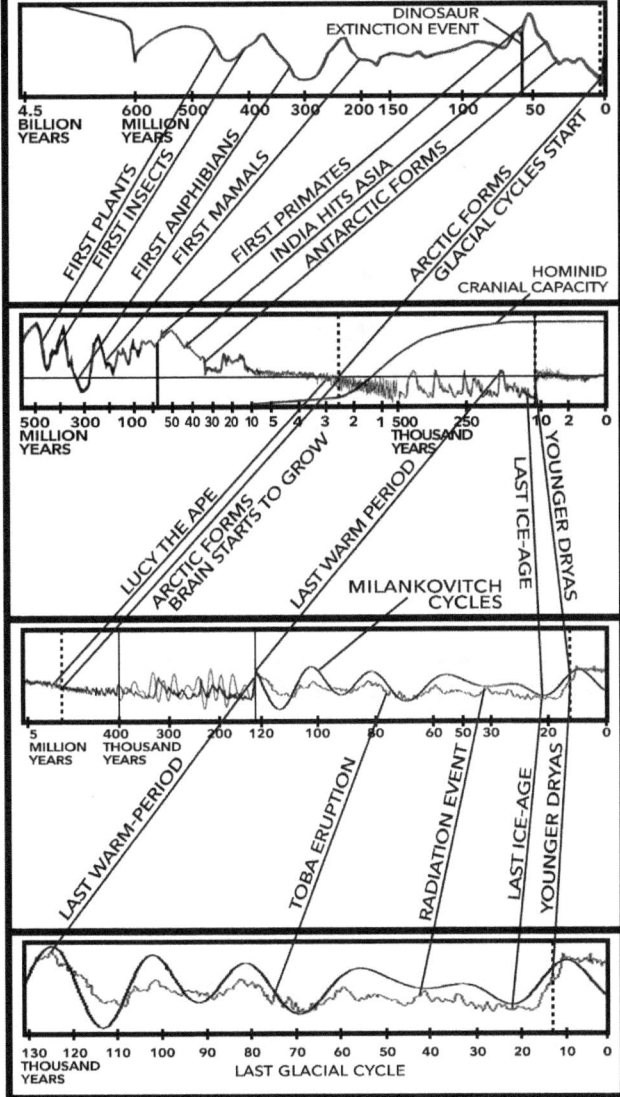

EARTH'S GLOBAL TEMPERATURE CHANGE

DINOSAUR
EXTINCTION EVENT

4.5
BILLION
YEARS

600 500 400 300 200 150 100 50 0
MILLION
YEARS

FIRST PLANTS
FIRST INSECTS
FIRST ANPHIBIANS
FIRST MAMALS
FIRST PRIMATES
INDIA HITS ASIA
ANTARCTIC FORMS
ARCTIC FORMS
GLACIAL CYCLES START

HOMINID
CRANIAL CAPACITY

500 300 100 50 40 30 20 10 5 4 3 2 1 500 250 10 2 0
MILLION   THOUSAND
YEARS   YEARS

LUCY THE APE
ARCTIC FORMS
BRAIN STARTS TO GROW
LAST WARM PERIOD
LAST ICE-AGE
YOUNGER DRYAS

MILANKOVITCH
CYCLES

5   400   300 200 120 100   80   60 50 30   20   0
MILLION THOUSAND
YEARS   YEARS

LAST WARM-PERIOD
TOBA ERUPTION
RADIATION EVENT
LAST ICE-AGE
YOUNGER DRYAS

130 120 110 100 90 80 70 60 50 40 30 20 10 0
THOUSAND
YEARS   LAST GLACIAL CYCLE

11

PASSING TIME

It is very hard for us to understand long periods of time. One-million years is so incredibly long, it's as-good-as infinite. Even one glacial cycle of one-hundred-thousand years is very hard to grasp.

One way to make sense of long-periods of time is by observing the mechanics of our Solar System. Milankovitch cycles give a great mathematical model of past and future global temperatures in regards to Earth's precession, tilt and eccentricity.

When the system is stretched-out, a state of high-eccentricity, strong gravitational forces put extra stress on all the planets. The stretching and compressing will crack Earth's crust, triggering mega-earthquakes and mega-volcanic eruptions. These forces can also pull-in comets from the outer solar system, increasing chances of asteroid impacts. These periods are rattled with disasters. Ever since life crawled-out of water some 470 million years ago, it has been "reset" many times by these cataclysms. But the rest of the time, Earth harbours long periods of stable global climate.

Sixty-six million years ago, a massive asteroid impacted the Yucatan Peninsula causing the extinction of the dinosaurs. This is an important event in our own evolution as it allowed the surviving mammals to dominate the planet. Without this disaster, there would be no primates for us to evolve from.

Evolution is hence in-sync with these large cataclysms, that are also in-sync with global temperature changes, that are also in-sync with Earth's movement.

This correlation is often represented with precessional numbers. They are the proportions within our Universe and its fractal-like behaviour.

The Great Pyramid was built adhering to these proportions. The Moon perfectly eclipsing the Sun, the Vitruvian Man, the human stride, the average life-span of a human, squaring of the circle, polygonal mathematics, the arrival of the comets, the glacial cycles and the flooding of Sahara. All bound by Earth's movement through space.

Our Solar System is currently in a relatively stable circular-state, with a moderate tilt.

We reached present-day temperatures 11,600 years ago, that put us into in a warm-glacial-period. Last such period ended 120,000 years ago.

The coldest-point happened 22,000 years ago, and the next ice-age will happen in about 55,000 years from today.

Glacial cycles have stabilized at about 100,000 years-in-length, and are a great way to reference and measure time.

Out of countless cataclysms that occurred during the last glacial cycle, certain ones dwarf them all.

The Toba eruption 74,000 years ago, caused a global volcanic-winter lasting six to ten years. It had a thousand-year-long cooling-episode and is linked to a genetic bottleneck in human evolution.

42,000 years ago, Earth's magnetic-field broke-down temporarily, and the intense Sun's radiation caused mass-extinctions and the demise of the Neanderthals.

The most recent event, called the Younger Dryas period, was the mother of all disasters. A twelve-hundred-year apocalypse, 12,800 to 11,600 years ago, is said to be the largest cataclysmic event in the past five-million years.

These disasters create pockets-of-time of about 30,000 years, with relatively undisturbed environment, where flora and fauna can take their time to evolve.

THE RHYTHM OF OUR UNIVERSE

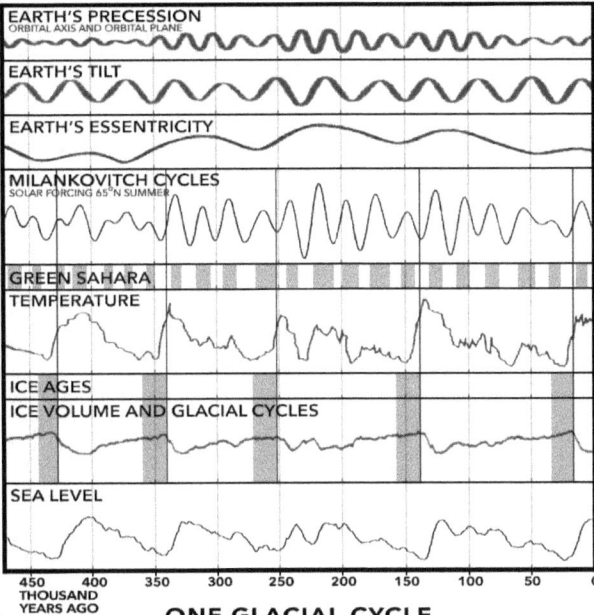

EARTH'S PRECESSION
ORBITAL AXIS AND ORBITAL PLANE

EARTH'S TILT

EARTH'S ESSENTRICITY

MILANKOVITCH CYCLES
SOLAR FORCING 65°N SUMMER

GREEN SAHARA

TEMPERATURE

ICE AGES
ICE VOLUME AND GLACIAL CYCLES

SEA LEVEL

| 450 | 400 | 350 | 300 | 250 | 200 | 150 | 100 | 50 | 0 |

THOUSAND
YEARS AGO

ONE GLACIAL CYCLE

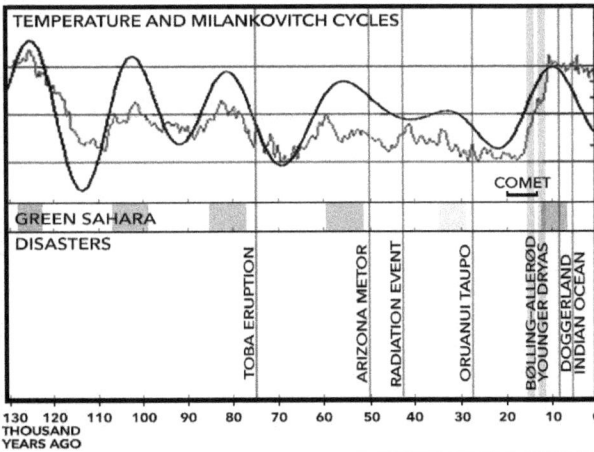

TEMPERATURE AND MILANKOVITCH CYCLES

COMET

GREEN SAHARA

DISASTERS

TOBA ERUPTION

ARIZONA METOR

RADIATION EVENT

ORUANUI TAUPO

BØLLING–ALLERØD
YOUNGER DRYAS

DOGGERLAND
INDIAN OCEAN

| 130 | 120 | 110 | 100 | 90 | 80 | 70 | 60 | 50 | 40 | 30 | 20 | 10 | 0 |

THOUSAND
YEARS AGO

HOW LITTLE WE KNOW

Given enough time, the four elements earth, water, air and fire will destroy everything they get in touch with.

Vegetation is the first thing that will start chewing away our cities and any trace we leave behind.

A metal car, unless somehow preserved, will completely disintegrate in a few-hundred years. Even plastic will dissolve into tiny micro-particles and disappear.

Best chance of preserving something would be to bury it. Though erosion, earthquakes or its discovery can eventually expose it.

People can also change, assimilate and sometimes displace, ravage or destroy previous cultures.

All this can cloud, puzzle and confuse what we might think had happened in our past.

Humans also tend to live close to water, the very place that erodes the most, and the coastlines where most of our ancient ancestors used to live, today deep under water.

There is very little left for us to discover.

The greatest testimony of how little we know about our past ancestors is the discovery of the hominid species called the Denisovans.

Only one finger in a cave in Siberia, and some teeth half-way across the world, in the mountains of Tibet have so far been discovered.

Yet this species was spread all across Australasia, with people of New Guinea carrying as much as eight percent of their DNA.

If we didn't stumble upon these few bones, we would have never known that Denisovans ever even existed.

This begs the question of how many other hominid species could have existed in our past, that we just haven't discovered yet, and maybe never will.

Most ancient discoveries are made purely by chance and often preserved under miraculous circumstances. It's why they are often found in some deep caves.

Earth is constantly hit by disasters. They are just as hard to discover, and the ones that we do discover are often ambiguous.

The strongest current hypotheses in what caused the Younger Dryas cataclysm is an asteroid impact from a comet that entered our solar system about 20,000 years ago.

One-hungered miles in diameter, it split into eight pieces, with the biggest one striking Greenland at the Hiawatha crater.

The remaining asteroids and the debris swept across Europe, Asia and the Middle East, leaving deposition evidence of ten million tons of impact spherules across four continents.

In our recorded history, there has really been no event that would be considered a real mega-catastrophe.

Possibly the biggest calamity since the Younger Dryas is the asteroid impact that might have struck the floor of the Indian Ocean about 5,000 years ago.

Hidden deep under water, at the Burckle crater, this impact would have unleashed a mega-tsunami of an unimaginable size.

Any reference to floods in the ancient texts and scriptures most probably refers to this event if not the Younger Dryas.

The reason why it is so hard for us to imagine an ancient world is because it is very hard, if not impossible.

There is one thing that can endure relentless disasters and long periods of time.

Large structures made of stone.

The pyramids.

It is exactly what they are, structures that can survive mega-cataclysms.

ANCIENT CIVILISATIONS

An archeological site in Turkey called Göbekli Tepe holds the key that unlocks our ancient past.

Only recently discovered, by accident, this megalithic site carbon-dates back to 11,600 years ago.

This date falls precisely at the end of the Younger Dryas period, so it is highly unlikely that the site was built during or at the end of this twelve-hundred-year cataclysm.

The site was actually buried by smaller stones, on purpose, and it is the burial that was carbon-dated. The actual construction of the site still remains unknown.

This discovery, its method of building, the matching stonework and the carvings, all unequivocally show that all megalithic structures around the world, including the pyramids were built by some ancient advanced civilization more than 12,800 years ago.

This civilization could have risen from the radiation event 42,000 years ago, and was then obliterated by the Younger Dryas disaster. A thirty-thousand-year reign.

Since Stonehenge and Göbekli Tepe are observatories, these people would have been highly in tune with the night sky and celestial objects.

They surely would have seen this comet, and in time realized that it was going to hit them. There really is no greater motive for building pyramids than this, even though they come across as some kind of giant power plants.

This ancient world would have had different cultures and beliefs. They would have had relationships with other hominid species and the mega-fauna, and would have lived through an ice-age. What was their language like? Their political structure? Fashion and food? Have they been to the Moon? Have they decoded the DNA? Could they control weather? Melt granite? Were they immortal? Were they even Homo Sapiens?

This ancient civilization would also not have been the only civilization that existed in our past.

A very recent discovery in Africa found two large wooden beams, waterlogged under a riverbed, dating back to about 476,000 years ago.

This finding has the scientists puzzled as it shows that hominids have known how to build fortified wooden structures for at least five glacial cycles. Fortification is the first step to having agriculture and growing large societies.

Since then, more than a dozen different ancient civilizations could have existed on Earth, separated by mega-disasters, all most likely unaware of each-other's existence.

Technology, industrial age and electricity have been around in the modern world for only just over a hundred years. Imagine a society that has had electricity for ten-thousand years.

What kind of trace do you think our civilization will leave twenty-thousand years from today, when another mega-calamity strikes? Will we survive it? Perhaps another hominid species will branch-out in the mean time?

It is an absolute miracle that the Sapiens survived the Younger Dryas period.

It brought the human race back to living as hunter-gatherers. Back to survival mode and the start of our own civilization.

A very young civilization.

We are one of many civilizations that have, and will continue to live on this planet, in various pockets of time, separated by major global disasters.

Any empire or kingdom in our history that is associated with megalithic sites, would have adopted, changed or built on top of previously founded structures.

ESTABLISHING A TIMELINE

Since our planet is slowly precessing, like a spinning-top, any star-alignments will evenly repeat with an interval of about twenty-six thousand years. This period is called a Great Year. The rate of this precession equals 1 degree on the horizon every 72 years. One full circle of 360 degrees, equals 25,920 years, which is one Great Year.

"Precession of the equinox" acts as a clock, passing through twelve zodiac constellations, where the handle is the eyesight of an observer looking east on the day of the spring equinox.

This rate gives the precessional numbers.
12, 36, 72, 108, 360, 2160, 25,920, 43,200.

All ancient megalithic sites reflect the night sky in some way, and aligning them accordingly will indicate certain times in the past.

It is often remarked how precisely the sides of the Egyptian pyramids are aligned with True North, the rotational axis of our planet. The Great Pyramid is off by only about three arc-minutes, one arc-minute being one-sixtieth of one degree, with the Moon as seen in the sky stretching about 31 arc minutes.

Considering the incredible precision of the pyramids and how they were built, a much more compelling argument is just the opposite.

With the two sides of the Great pyramid almost parallel to each other. East -3.4' and West -3.7' arc minutes. The builders could have aligned the pyramids within one arc minute of the exact True North, or any other deviation they liked.

The African tectonic plate is colliding with Eurasia and rotating counter-clockwise at that very location, about one arc-minute every 4,630 years.

If the builders had aligned the Great Pyramid with True North, the construction time of the pyramids calculates to about sixteen-and-a-half thousand years ago. The onset of a global glacial meltdown called Bølling-Allerød, as Sahara was entering its last fertile period.

Looking at cardinal deviations, the overall averages of the seven largest and most preserved Egyptian pyramids, the range of the data error is really only consistent with the scale of the Milankovitch cycles or the Great Year, it is too large for zodiac.

The smallest of the three Giza pyramids is the only one that points right of True North, far into the future. Over 12 arch-minutes, it seems more of a clue than a large error. This point in time is about 57,000 years from today, roughly the next ice-age.

The Second of the Giza pyramid points to about 24,000 years ago, around the previous ice-age.

The three pyramids, Bent, Meidum and Sahure all match the local global temperature maximums of the past glacial cycle, that also matches the times when Sahara was green.

The Red pyramid matches the time of a cataclysm. Being red in colour, perhaps it was meant to point to this disaster. If could also represent the start of their own civilization.

The builders of the Egyptian pyramids would have known that the African plate was rotating and they could have used the deviations to create a time-map.

It is also quite possible, that people that built these pyramids were not the first people to build structures aligned to True North.

Pyramids were often built on top of previous pyramids, and doing so would commemorate the original cardinal deviations of the structures below.

Most of these rotations could of course have been caused by earthquakes, especially the smaller pyramids with no casing stones, making all this data completely random.

However, three particular sides of the Great pyramid's casing stone deviations all fall within one arc minute range. And the same is true with the second Giza pyramid.
Great -3.6N -3.4E -3.7W: -3.57 = 16,529 years
Second -3.8N -4.0E -4.2W: -4.00 = 18,520 years
Average -3.78 = 17,501 years

This data shows not only that the Giza plateau remails stable, but that the builders could align them with incredible precision. The average of this data gives us the best guess for when the pyramids were built, seventeen and a half thousand years ago.

This time would explain why there are no pyramids in the far north, where the glaciers were at the time.

Stonehenge and Serpent Hill happen to be located right at the edge of the maximum glacial advance during the last ice-age.

Weather its due to its casing stones being removed earlier or the granite-clad casing that's different from the polished limestone of the first two, the third pyramid's deviation still remains a mystery.

PYRAMID CARDINAL DEVIATIONS - A TIME-MAP

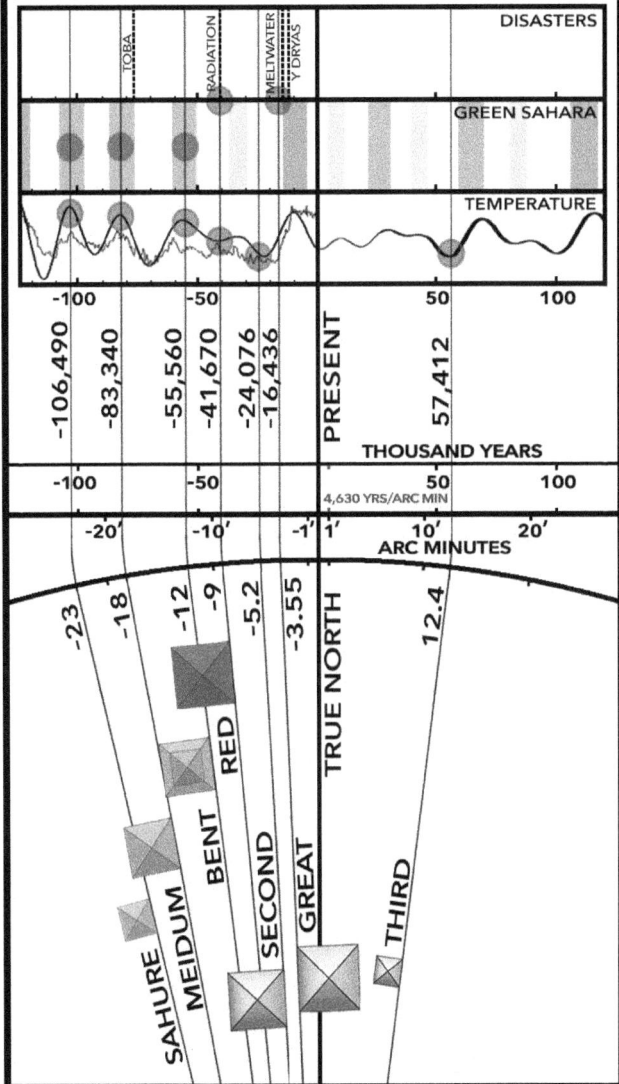

WE CAN DO BETTER

There is one thing that can survive the wrath of time even longer than the pyramids.

It is life itself.

Life will evolve in similar ways on any Earth-like planet in the universe. The physical properties of water, gravity and air dictate it.

A dinosaur called Ichthyosaurs was an air-breathing marine reptile that very closely resembled a dolphin.

A mammal and a reptile went back into water independently, and evolved into almost identical shapes. This is not a coincidence.

Any world with our physical properties will have similar evolutionary outcomes. The most efficient kinematic movement will most probably produce something very similar to a cat. Likewise, the properties of sound and vocalization will lead to development of languages alike.

There is a part in our brain that changed more than a million years ago that sets us apart from animals. We can't ignore the fact that we are animals ourselves.

There is no animal in the wilderness that does not utilize their full physical and mental capacity. They would not survive. They have to be the best version of themselves, at all times.

The necessity for hard physical work is hence engraved into us, though our evolution.

Yet, the fact that our bodies are not protected by hair and that we can't digest raw meat, makes it obvious that we evolved into hominid species with sheltered and comfortable lives.

Perhaps an evolutionary advancement, but excess comfort can lead to lethargic, unhealthy and allopathic lifestyles.

Regardless of comfort, when a mega-cataclysm hits, everything changes. People scatter and re-group into small communities with a simple and basic need to survive. Maslow's hierarchy of needs becomes very relevant.

Though such strife can easily evoke savage behaviour, in most cases a stranded survivor, upon arrival would receive help.
Food and shelter, the basic sustenance.

This person would then give back to the community with their hard work, skill and ambition, and grow as an individual for self-actualisation.

Through time, they would all rejoice in the community itself, the most important part of life, the human connection.

This empathic paradigm is the real-world rock-bottom model for any society, harbouring unified, healthy and content human beings.

In time however, societies change, and if they fail to provide its people with means of acquiring basic sustenance, people will turn desperate. They will turn to crime.

To create a content human being, a community, an organization or a society, the three human needs, sustenance, community and self-actualization, need to all be in balance.

The ancient symbol of the crook and the flail was perhaps meant to represent the importance of this essential human need in maintaining a healthy society.

Our current global civilization is not the first nor the last to live on this planet, and with all its advancements and achievements, in its seven-thousand-year existence, it has never seen time of world-wide peace.

A civilization with no crime.
A strong, meek and caring society.

Serpents of Knowledge

9 781738 200009

www.ingramcontent.com/pod-product-compliance
Lightning Source LLC
Chambersburg PA
CBHW070948210326
41520CB00021B/7104